MANTRAS WITH SPIRITUAL CONCEPTION OF GOD IN A CIRCLE: ORIGINAL SANSKRIT MANTRAS WITH ENGLISH TO FIX ANY PROBLEM, FOR INNER HAPPINESS, ENLIGHTENMENT AND HIGHEST STATE OF ECSTASY

Copy Right BLGS LLC

All rights reserved. No part of this may be used or reproduced in any form or by any means, or stored in a database or retrieval system, or transmitted or distributed in any form by any means, electronic, mechanical photocopying, recording or otherwise, without the prior written permission of author or publisher. The information provided is for only instructional value. This book is sold as is, without warranty of any kind, either expresses or implied. This e-book/Book is provided "as is" without warranty of any kind, either express or implied, including, but not limited to, the implied warranties of merchantability, fitness for a particular purpose, or non-infringement. In no event shall the authors or copyright holders, publisher, distributor be liable for any claim, damages or other liability, whether in an action of contract, tort or otherwise, arising from, out of or in connection with the book or the use or other dealings in the book.

Why This Book:

Vedic Sanskrit sacred writings contains over 1000s of chapters and written in a complex language, people are very busy with life and do not have enough time to spend to read entire Vedic books. Here we are providing a Mantra taken from Kali-Santarana Upanishad, a Mantra from Uddhava Gita and a Rigveda Mantra With Spiritual Conception Of God In A Circle, Krishna Beej Mantra to fix any problem, for inner happiness, enlightenment and highest state of ecstasy.

Most of the books give you the transliteration of Mantra and Shlokas in English only.

We have provided Sanskrit words in Devanagari script (a syllabic script used in writing Sanskrit) and transliterated into Roman script. It is a well-known scientific fact that Sanskrit Mantras have healing effects and Neuroscience confirms "The Sanskrit effect." These Mantras encapsulate all the power of the cosmos in them to fully satisfy one's spiritual needs. Now written in English, will help Every human being including people in English speaking countries. These Sanskrit Mantras are for all human beings. In Vedic religion, Vedic Sanskrit was considered the language of the gods. A Sanskrit word represents sound of the desired object.

Devanagari =Deva (god) + Nagari (city) = City of the Gods

Symbolic meaning of the city is the body itself also there is multiple layers of symbolism associated with each Word and sound. Symbolic meaning of the city is the body itself also there is multiple layers of symbolism associated with each Word and sound. So, when you Chant or meditates on the specific sounds of the Devanagari alphabet, the written form also appears in the mind. Original Sanskrit Text with English for Healing. This is for the benefit of all humanity because these Mantras encapsulate all the power of the cosmos in them. Please take two minutes from your busy life to enjoy this Vedic goodness and read this for meaningful daily life with favorable results.

MANTRA IN DIVINE LANGUAGE SANSKRIT:

ORIGINAL SANSKRIT TEXT

ENGLISH TRANSLITERATION

HARE KṚṢNA HARE KṚṢNA KṚṢNA KṚṢNA HARE HARE HARE RĀMA HARE RĀMA RĀMA RĀMA HARE HARE

ENGLISH TRANSLATION

ORIGINAL SANSKRIT TEXT

ENGLISH TRANSLITERATION

ENGLISH TRANSLATION

ORIGINAL SANSKRIT TEXT

ENGLISH TRANSLITERATION

ENGLISH TRANSLATION

ORIGINAL SANSKRIT TEXT

KRISHNA BEEJ MANTRA

ENGLISH TRANSLITERATION

ENGLISH TRANSLATION

MANTRA TO INVOKE THE PROTECTION OF LORD VISHNU

MEDITATION STEPS

GOD & MANTRA

VISUALIZE GOD:

VISUALIZE LORD VISHNU

TIME & PLACE

SEAT & SITTING POSTURE

RELAX & CHANT.

BENEFITS OF RUDRAKSHA JAPA MALA

WORSHIP & CONCLUDE.

Mantra in Divine Language Sanskrit:

These Mantras were composed by the ancient Vedic saints in the divine energy based language **of Sanskrit and hence these mantra produce powerful** energy-based sound with specific spiritual potentialities or power or specific energies related to health, wealth, happiness, healing, prosperity, inner peace, love, protection, luck, illumination, happiness contained within the vibration of the word is realized within us, so These Mantras are POWERFUL, SANSKRIT AFFIRMATIONS ,words of Power, Divine Power transmitted through words.

We have provided Sanskrit words in Devanagari script (a syllabic script used in writing Sanskrit) and transliterated into Roman script, so it is much easier to pronounce these mantras. These affirmations Mantras must be chanted in Sanskrit to stimulate the positive energy related to the objective you need to accomplish. In Vedic religion, Vedic Sanskrit was considered the language of the gods. A Sanskrit word represents sound of the desired object, so you need to perform Sanskrit Mantra Japa, i.e., repeated rhythmic chanting, repetition of the mantra. Tantra means a method and Yantra is a geometric figure representing an aspect of divinity or higher state of consciousness. So, you can use this e-book for Mantra, Mantra and Yantra.

Devanagari =Deva (god) + Nagari (city) = City of the Gods

Symbolic meaning of the city is the body itself also there is multiple layers of symbolism associated with each Word and sound. Symbolic meaning of the city is the body itself also there is multiple layers of symbolism associated with each Word and sound.
So when you Chant or meditates on the specific sounds of the Devanagari alphabet, the written form also appears in the mind. Sanskrit Mantras are pure vibration sound representing God so it's important to fully read and understand hidden meanings.
These Mantras were composed by the ancient Vedic saints in the divine energy based language of Sanskrit and hence these mantra produce powerful energy-based sound with specific spiritual potentialities or power or specific energies related to health, wealth, happiness, healing, prosperity, inner peace, love, protection, luck, illumination, happiness contained within the vibration of the word is realized within us,

so These Mantras are powerful, Sanskrit affirmations, words of Power, Divine Power transmitted through words. Collection of Most Powerful mantras along with their English translation.

Most sacred and oldest available Divine hymns, Sanskrit Mantra. It is hoped that the devotees will use these mantras with full understanding and devotion to fulfill their goals, Health, wealth, happiness, success, Abundance, Well Balanced Life. All aspects of life.

Sanskrit MANTRAS are pure vibration sound representing God, SO IT IS IMPORTANT TO FULLY READ AND UNDERSTAND HIDDEN MEANINGS FOR EXAMPLE AS PER CHANDOGYA UPANISHAD.

OM IS THE ESSENCE OF THE ENTIRE UNIVERSE,

THE ESSENCE OF ALL BEINGS IS THE EARTH.

THE ESSENCE OF THE EARTH IS WATER.
THE ESSENCE OF WATER IS THE PLANT.
THE ESSENCE OF THE PLANT IS MAN.
THE ESSENCE OF MAN IS SPEECH.
THE ESSENCE OF SPEECH IS THE RIG-VEDA.
THE ESSENCE OF RIG-VEDA IS THE SAMAVEDA.
THE ESSENCE OF SAMAVEDA IS OM.

A Sanskrit word represents sound of the desired object, so you need to perform Sanskrit Mantra Japa, i.e., repeated rhythmic chanting, repetition of the mantra.

Maha-Mantra: This 16-word Maha-Mantra is taken from Kali-Santarana Upanishad.

Kali Yuga is associated with the demon Kali, it is an age of quarrel and hypocrisy, Many diseases, age of misery, Ignorance of dharma.

Sage Narada Messenger of Gods and enlightening wisdom revealed that these sixteen names destroy the negative effects of Kali and helps to overcome the effects of Kali Yug. These sixteen names destroy the 16 envelopments in which the jiva is enveloped and Only Parabrahman is revealed. Jiva, after annihilation of Sixteen Kala, becomes Brahman itself, it unites with Brahman.

16 Avarana in Prashna Upanishada:
1.PrAna = life force, 2.Shraddha = faith, 3.kham = ether, 4. VAyu = air, 5.Jyoti = Light, 6.Apah = water, 7.Pruthwi= earth, 8. Indriya = sense organs, 9. Manah= mind, 10. Annam = food eaten, 11. Viryam = vital energy, 12. Tapah = Knowledge, 13Mantrah = sonic power, 14. karma = actions and their reactions, 15. lokah = the realms of existence, and 16.NAma = Individuation

16 names to destroy 16 envelopments:

1.Hare, 2. Krishna, 3.Hare, 4.Krishna, 5.Krishna,

6.Krishna, 7.Hare, 8.Hare,

9.Hare, 10.Rāma, 11.Hare, 12.Rāma, 13. Rāma,

14.Rāma, 15.Hare, 16.Hare

Original Sanskrit Text

हरे कृष्ण हरे कृष्ण , कृष्ण कृष्ण हरे हरे ।
हरे राम हरे राम , राम राम हरे हरे ॥

English Transliteration

Hare Kṛṣṇa Hare Kṛṣṇa
Kṛṣṇa Kṛṣṇa Hare Hare
Hare Rāma Hare Rāma
Rāma Rāma Hare Hare

English Translation

Oh Lord, Oh energy of the Lord, All Attractive One, O Rama, Reservoir of Pleasure, please engage me in Your service.

Benefits : **Remove anxiety and inner fears, remove bad habits, evil thoughts & Negative emotions, achive highest state of ecstasy joy, peace and happiness. It relieves from all miseries, provides divine Love, mercy and inner happiness. It purifies yourself and surrounding, It cleans your mind and heart. Fix Kali Yug Problems.**

Uddhava Gita Ras Lila Mantra
Uddhava Gita is the spiritual instructions of Krishna to Uddhava, spread over 23 chapters comprising more than 1000 'verses'. Here We are providing the main mantra.
Original Sanskrit Text

या वै श्रियार्चितमजादिभिराप्तकामैर्
योगेश्वरैरपि यदात्मनि रासगोष्ठ्याम् ।
कृष्णस्य तद्भगवतश्चरणारविन्दं
न्यस्तं स्तनेषु विजहुः परिरभ्य तापम्

English Transliteration

yā vai śriyārcitam ajādibhir āpta-kāmair
yogeśvarair api yad ātmani rāsa-goṣṭhyām
kṛṣṇasya tad bhagavataś caraṇāravindaṁ
nyastaṁ staneṣu vijahuḥ parirabhya tāpam

English Translation

At the time of Ras Leela, Lord Krishna's dance of divine love was so divine and transcendental that it gives the spiritual bliss to all devotees. These Gopis took the lotus feet of the Supreme Lord Krishna, who is worshiped by Goddess Lakshmi, Brahma and other deities, and by great yogis who are free from all desires and embraced them, prostrates unto himself by holding his feet., and thus were freed from their sorrows.

Rigveda Mantra With Spiritual
CONCEPTION OF GOD IN A Circle
The Rigveda is one of the four sacred canonical texts is a collection of 10 books with 1,028 hymns in about 10,600 verses. Here Providing you a RIG VEDA mantra with the technical translation that gives the VALUE OF PI .

Original Sanskrit Text

गोपीभाग्य मधुव्रातः
श्रुंगशोदधि संधगिः |
खलजीवतिखाताव गलहाला
रसंधरः ||

English Transliteration

gopeebhaagya maDhuvraathaH
shruMgashodhaDhi saMDhigaH

khalajeevithakhaathaa- va galahaalaa
rasaMDharaH

English Translation

Oh Lord Krishna, anointed with curd from the milk of worship O Savior of the fallen, O Lord of Shiva, please protect me.

Consonant Code:

ka (क) – 1, kha (ख) – 2, ga (ग) – 3, gha (घ) – 4, gna (ङ) – 5, cha (च) – 6, cha (छ) – 7, Ja (ज) – 8, Jha (झ) – 9

ta (ट) – 1, tha (ठ) – 2, da (ड) – 3, dha (ढ) – 4, ~na (ण) – 5, Ta (त) – 6, Tha (थ) – 7, Da (द) – 8, Dha (ध) – 9

pa (प) – 1, pha (फ) – 2, ba (ब) – 3, bha (भ) – 4, ma (म) – 5

ya (य) – 1, ra (र) – 2, la (ल) – 3, va (व) – 4, Sa (श) – 5, sha (ष) – 6, sa (स) – 7, ha (ह) – 8

kshah (क्ष) – 0.

When we apply the Code to the Mantra:

gopeebhaagya maDhuvraathaH
shruMgashodhaDhi saMDhigaH

khalajeevithakhaathaa- va galahaalaa
rasaMDharaH

It Gives value of PI
=3.14159265358979323846 2643- 3832792...

A circle is a shape consisting of all points in a plane that are at a given distance from a given point, the center. The distance between any point of the circle and the center is called the radius.

गोपीभाग्य मधुव्रातः श्रृंगशोदधि संधिगः |
खलजीवितखाताव गलहाला रसंधरः ||

gopeebhaagya maDhuvraathaH shruMgashodhaDhi saMDhigaH

khalajeevithakhaathaa- va galahaalaa rasaMDharaH

PI =3.14159265358979323846264 3- 3832792...

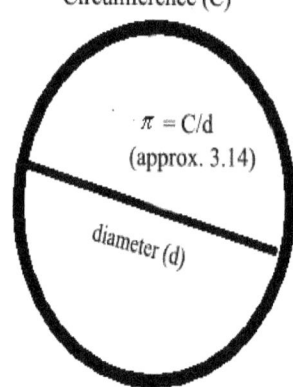

Pi (π) is commonly defined as the ratio of a circle's circumference C to its diameter D.

The circumference is the perimeter of a circle.

The ratio C/D is constant, regardless of the size of Circle. Pi remains the same. Pi reveals that everything small or big is surrounded (circumference) by Gods Equal Supreme Love, goes straight through the central point (diameter) and is infinite.

Original Sanskrit Text

KRISHNA BEEJ MANTRA

कृं कृष्णाय नमः

English Transliteration

Om Klim KRISHNAYA NAMAHA

English Translation

This mantra is the root mantra of Lord Krishna. Kleem is the principle of attraction, thus attracting the energies of Krishna in your life to connects us to that higher energy and helps us to break away from that endless cycle and attachment.

Mantra to invoke the Protection of Lord Vishnu

Original Sanskrit Text

ॐ विष्णोर्नुकं वीर्याणि प्रवोचं यः पार्थिवानि विममे रजाँसि यो अस्कभायदुत्तरँ सदस्थं विचक्रमाणस्तेधोरुगायो

English Transliteration

Vishnornukam veeryani pravocham ya parthivani vimame rajaasi |
yo askabhaya duthara Sadastham vichakramanas threthorukayo ||

English translation

We will sing the noble deeds
of that Vishnu:
who measured dust mites
Who is the protector and
protector of the universe
One who measured the three
worlds in three steps.
Who is praised by great sages
and devotees

Meditation Steps

Repetition of a Mantra is the basis of Mantra Meditation.

You may contact a spiritual teacher.

Mantra Meditation is the easiest and safest form of meditation and here are the steps:

Sit calmly on a Kusha asana.

Choose a Mantra (Given Below).

Close your eyes and try to keep your mind fixed on the meaning of the mantra during the meditation practice.

You may chant the mantra aloud.

God & Mantra

Select your Mantra and Istadevata: Ishta Devata means the preferred form of God. According to Bhakti yoga, meditation on any aspects of God leads to enlightenment, realization, ultimately resulting in moksha so choose the Mantra with Ishta Devata. Devotees can invoke God in whatever form a devotee prefers.

Visualize God:

Lord Krisna is depicted with blue skin, wearing a peacock-feather crown, and playing a flute.

Visualize Lord Vishnu

Vishnu appears as a young, very handsome man, with four arms in which always carry four objects

1. The conch: the sound this produces 'Om', represents the primeval sound of creation
2. The chakra, or discus: mind
3. The lotus flower: cosmic harmony, existence and liberation
4. The mace: mental and physical strength

Time & Place

You can select a time that fits your schedule best but as per Vedic sage the best time to meditate is in the morning hour before dawn.
Choose a quiet place for your meditation in your home where you are unlikely to be disturbed.

Seat & Sitting Posture

Spread the meditation Seat and sit upon it in a lotus position it is a cross-legged sitting posture to allow the body to be held completely steady as the body is steadied the mind becomes calm.

Kusha Asana is recommended in Vedas, and the Bhagavad-Gita Gita. Kusha is the root for the Sanskrit word for "expert," Kosala.
Please consult your health care provider before trying any kind of meditation.

Relax & Chant.

Now think of the infinite sky and relax. Our body has nine openings or gates i.e., the two eyes, two ears, two nostrils, one mouth & two lower openings, withdraw your consciousness from all these gates into heart and imagine and pray to your God (Istadevata), Concentrate your mind on the Istadevata for up to 10 minutes or a time suitable to you.

It is recommended that for Fulfillment of all one's desires a mantra should be chanted 108 times. According to astronomy, there are 12 signs of zodiac and 9 planets in the universe and if multiplied, 12x9=108. Thus 108 are considered to unite the Individual consciousness with the absolute consciousness. And hence a Japa Mala usually consists of a string of 108 beads, with one summit bead known as a sumeru. A mala / rosary would always have an extra bead (109th) as the extension to the row of beads, whose total number is usually 108. This 109th bead is called "Sumeru" is a static point in these malas as it marks the beginning & end of the rosary. It is hoped that the devotees will use these mantras with full understanding and devotion to fulfill their goals, Health, wealth, happiness, success Chanting these mantras with faith and devotion, will elevate your level of consciousness, and lead to higher spiritual attainment and enlightenment.

Benefits of Rudraksha Japa Mala

It has been said that people of all caste, country, creed and religion should use Rudraksha for peace, health, prosperity, physical, mental and spiritual well-being. The Tears of Lord Siva are called Rudraksha, are the original Vedic Beads of Power worn by the Gods, Yogis and Rishis since times immemorial on their path to Enlightenment and Liberation. History shows successful people adorning these rudraksha mala to gain power, health, protection and self-empowerment. Veda have mentioned that Rudraksha can align the positive energies around us to specific aspects of health, wealth happiness, spiritual fulfillment and enlightenment

Worship & Conclude.

Now give offerings via visualization through mudra, mental worship, and these items are all offered mentally by Panchopachara Pooja (5-Fold Worship)

Mentally Offer Gandham (sandalwood paste/powder),

Mentally Offer Pushpam (flower),

Mentally Offer Sugandham dhoop (incense),

Mentally Offer Jyotham (light) and

Mentally Offer Naivedyam (fruits and sweets with drinking water) respectively.

Compiled by Pandit Bharadwaj

Disclaimer:

Sanskrit devotional verse song written in the praise of God, and it was an oral tradition in Vedic period. Disclaimer: BLGS LLC HEREBY DISCLAIMS ALL WARRANTIES AND CONDITIONS; Descriptions for Mantras, products are taken from scripture, written and oral tradition. Original Text is made available under the Creative Commons License. Mantras are not intended to diagnose, treat, cure, or prevent any disease or condition. We make no claim of supernatural effects. All e-book items are sold as curios and entertainment only should not be used as a substitute for advice, programs, or treatment that you would normally receive from a licensed professional.

www.ingramcontent.com/pod-product-compliance
Lightning Source LLC
Chambersburg PA
CBHW070908220526
45466CB00005B/2177